# Electricity and Magnetism

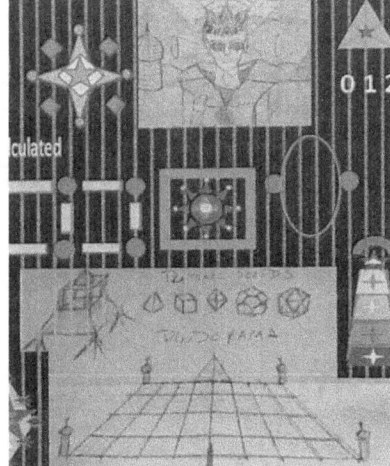

© 2022 Diogo de Souza
All Rights Reserved.

**Contact Information:**
diogodesouza7@gmail.com
diogodesouza7@hotmail.com

# My Theory of Quantum Pieces

Knowing that:

$E = mc^2$

The Planck's Energy h which is the smallest Energy possible can be used to find Planck's Mass the smallest Mass possible in the Universe:

$h = mc^2$

$m = \dfrac{h}{c^2}$ Planck's Mass

And since mass comes from energy:

$$\frac{mass\ of\ entire\ universe}{volume\ of\ universe} = p$$

Where p is the Density of the Universe.

The Planck's Length is the Smallest Length in the Universe. The Planck's Length cubed gives the Planck's Volume:

$h^3$ = Quantum Volume

Planck's Time is how long it takes the Speed of light to travel a Planck Length.

$$\frac{Planck's\ Length}{Speed\ of\ Light} = \text{Planck's Time}$$

$$\frac{volume\ of\ entire\ universe}{quantum\ volume} = U$$

U is how many Planck's Volume Cubes in the Universe.

$$\frac{Energy\ of\ entire\ Universe}{U} = h$$

Where h is Planck's Energy that fits inside a Planck's Volume.

Matter Universe has Energy.

Anti-Matter Universe has the same amount of Energy but negative.

The Volume of the Matter Universe that we see times two

gives the complete size of Matter plus Anti-Matter region in head on collision.

The Universe was created from Light that separated Matter from Anti-Matter and our Matter Universe will eventually collide back with the Anti-Matter Universe becoming just pure Energy in the form of Light.

There might be 10 Dimensions in the universe:

1 Up Down

2 Right Left

3 Forward Backward

4 Forward in time

5 Backward in time

6 Teleportation

7 Entangled Particles

# The next 3 are like video games:

8 Moving up appearing at the bottom. Moving down and appearing at the top.

9 Moving left and appearing at the right. Moving right and appearing at the left.

10 Moving forward and appearing backward. Moving backward and appearing forward.

# Only 4 of these Dimensions are big enough for us to see them. The other 6 are curled up into a very small space but Particles do experience them.

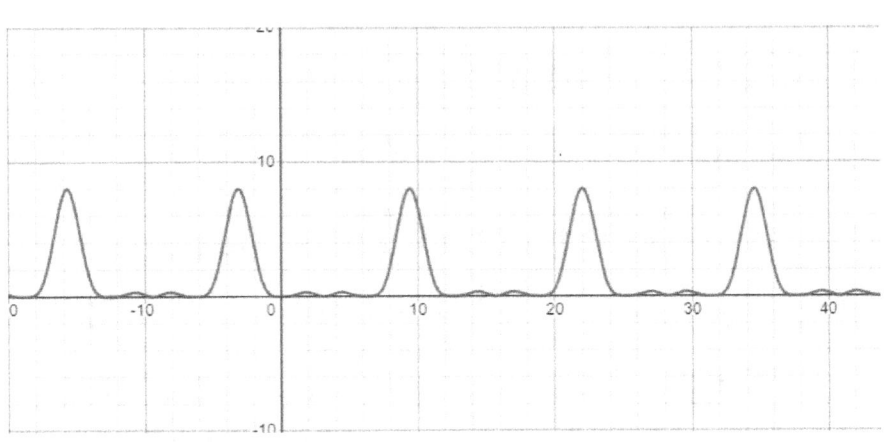

Quantum Fluctuations in Space lead to creation and annihilation of Universes through Time.

Universes exist only in the valleys not on the peaks.

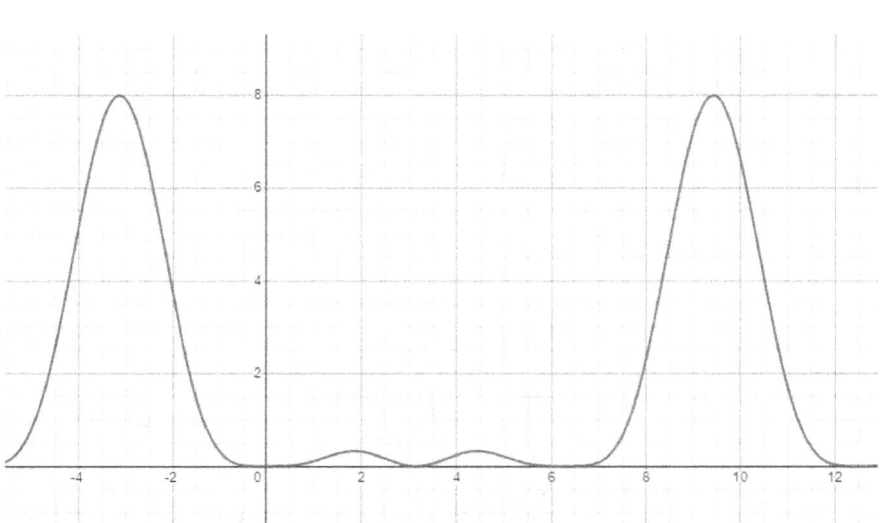

By zooming into these waves, we see two smaller peaks in the valleys which represent the expansion and contraction of a Universe.

The equation used as an example is:

$Y = (1-\sin(x/2)) \cdot (1-1\cos(x))\^2$

The x axis represents time and the y axis the volume of the Universe. The larger peaks are peaks of explosion. The Universes are only found in the valleys where the two smaller peaks are located.

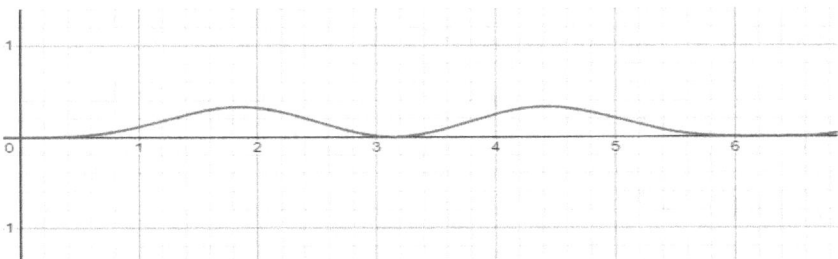

By zooming even closer to the two small peaks we see them better.

The bottom graph shows the Speed of the moving arrows leading to expansion and contraction of a Universe.

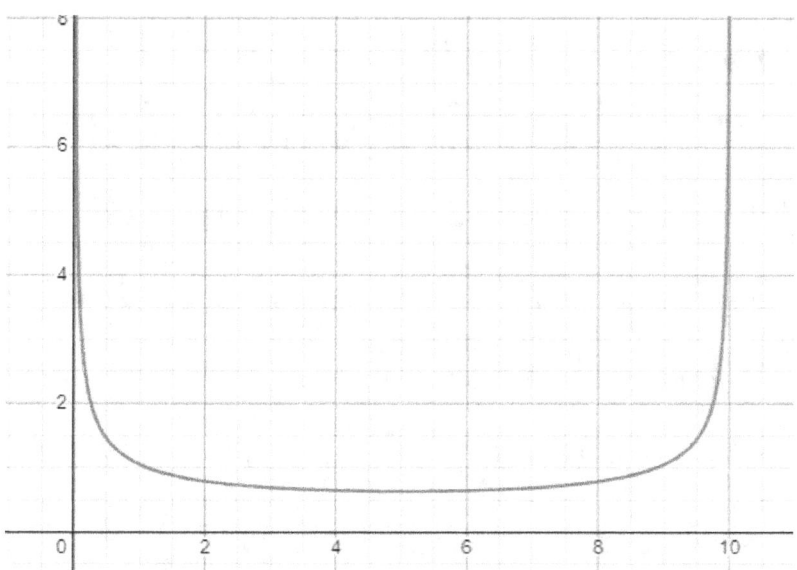

Near the beginning the Velocity of expansion is extremely large leading to Inflation Theory, and near the end the Velocity of contraction is also large. The graphs do not peak to infinity because the smallest size of a Universe can not be smaller than the Planck's Length. The graph peaks and then stops at the corners. In between the Big Bang and the Big Crunch the graphs becomes close to a flat line which means that the Velocity of expansion or contraction from moving arrows becomes stable and roughly constant.

The equation used in this example is:

$$\sqrt{\frac{1}{x} + \frac{1}{10-x}} = y$$

Where y is the Speed of the arrows and x are Arc Lengths that they make from the singularity.

In the next page there is a picture of these moving arrows through time:

# Stages of Cosmic Evolution:

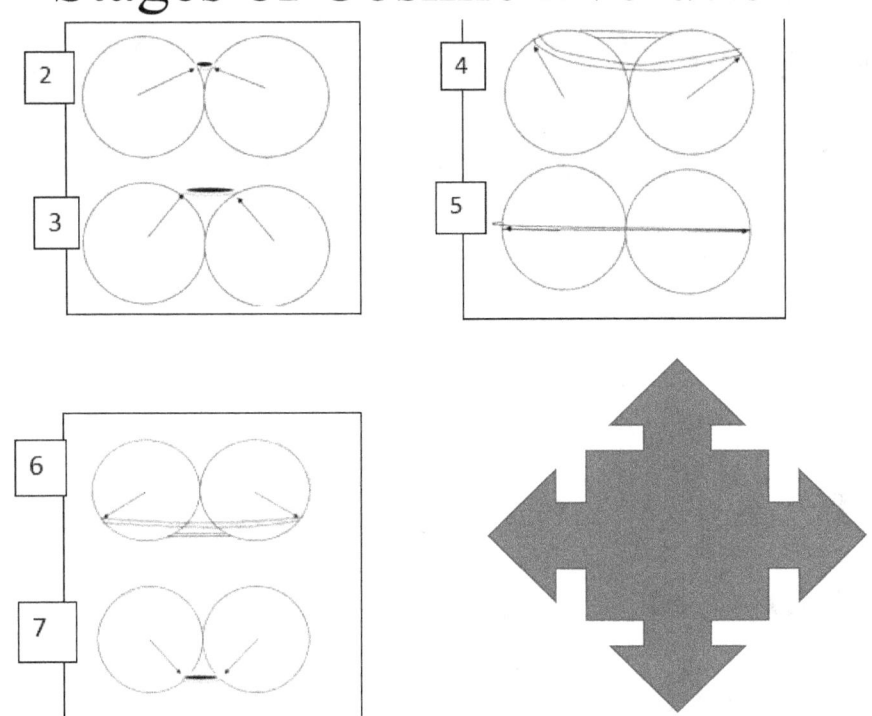

In the following page there is a table for the equation on page 13:

| Arc Length | Speed Of Arrows |
|---|---|
| 0.01 | 10.005004 |
| 0.1 | 3.1782086 |
| 0.2 | 2.2387698 |
| 0.5 | 1.4309525 |
| 1 | 1.0340926 |
| 2 | 0.79056942 |
| 3 | 0.69006556 |
| 4 | 0.64549722 |
| 5 | 0.63245553 |
| 6 | 0.64549722 |
| 7 | 0.69006556 |
| 8 | 0.79056942 |
| 9 | 1.0340926 |
| 9.5 | 1.4309525 |
| 9.9 | 3.1782086 |
| 9.99 | 10.005004 |

Near the very beginning and near the very end the Speed of the arrows leading to expansion and contraction are enormous:

| 0.0000001    | 3162.2777 |
| 0.0000000001 | 100000    |

| 9.9999999 | 3162.2777 |
| 9.999999  | 1000.0001 |

At time zero an enormous expansion occurred beyond Light Speed such as mentioned in Inflation Theory. In this example, the number 10 is used to represent the Circumference of the wheels in this model.

One complete revolution is Arc Length equal to 10 units.

Now here we are ready for the entire theory and an explanation of where these numbers come from:

# Math Wonders

© 2021 Diogo de Souza
All Rights Reserved.

**Contact Information:**
diogodesouza7@gmail.com
diogodesouza7@hotmail.com

# PINDORAMA

### Diogo's Universe

### Written Diogo Franklin de Souza

© 2017 Diogo de Souza
All Rights Reserved.

The first attempt to this theory began in 2005 while I was a student at a high school physics class. Over time a few modifications have been made but the overall idea and content did not change. 2018

I dedicate this book to the great nation of Israel.

## Introduction:

The origin of the matter in the universe are the many quantum fluctuations found throughout infinity giving birth to many other universes. There is no such thing as nothingness. Even in a vacuum, particles are created and destroyed several times. The matter in the universe in the like manner came from these fluctuations that will one day lead to their destruction in the Big Crunch.

Dark Matter is a combination of Rogue Planets, Brown Dwarfs, Higgs Bosons, gases, and Weakly Interacting Particles.

Without Dark Matter space would have no galaxies, stars, and planets. Dark Matter forms a structure in the universe allowing it to stay in place and exist. Dark Matter is also in great part, the Ether physicists referred to in the past.

Matter and Anti-Matter are created from nothing and destroyed after colliding with each other. All Matter in the universe was created likewise, and will eventually collide with the Anti-Matter in the Big Crunch. A new cycle will begin after that.

Dark Energy causes the expansion. Matter and Anti-Matter separate. Later the Dark Energy will cause the cosmos to contract leading Matter to collide with Anti-Matter, and that will be the end of a cycle.

The sum of all of the energy Matter plus Anti-Matter is exactly zero. $1 - 1 = 0$. From nothing, everything came to be.

The average density of the universe at the Planck Level is zero.

The universal sphere contains a doughnut inside with matter travelling along its sides an infinite number of times. There is an infinite number of cycles of creation and destruction.

# My Theory in Details

The universe is in the shape of a doughnut inside a sphere.

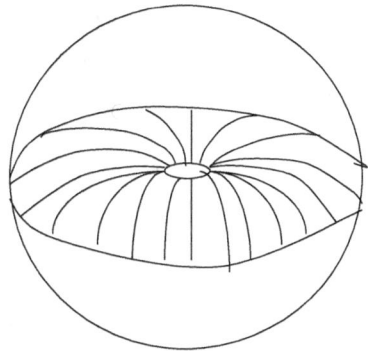

By cutting the doughnut in half and looking through one side we see two wheels. The space between the two circles is the Plank Length.

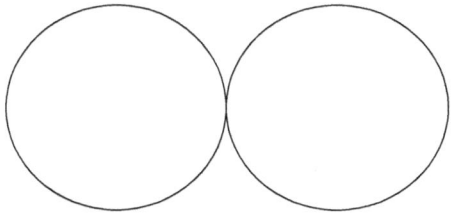

Cutting the doughnut in half as shown.

Imagine cutting the doughnut in half like as follow:
But in my theory the space between the two circles is almost zero.

In the center between the two wheels is the singularity. When the singularity containing all the matter in the cosmic bubble exploded, the expansion of that bubble began. The Matter and Anti-Matter in the universe expand separating from each other due to the Dark Energy caused by a gravitational dipole. $\bar{h} + R - R\cos(0) = \bar{h}$ is the distance between the tip of the arrows.

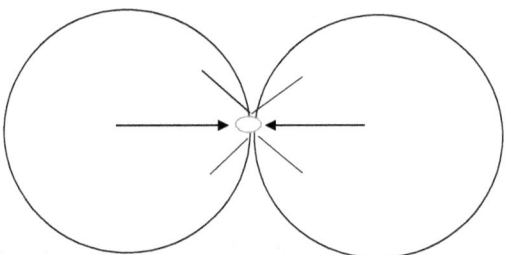

> The Planck's Constant above is for length. It is the smallest length possible for the universe. It is how wide was the singularity before the Big Bang.

The arrows indicate position.

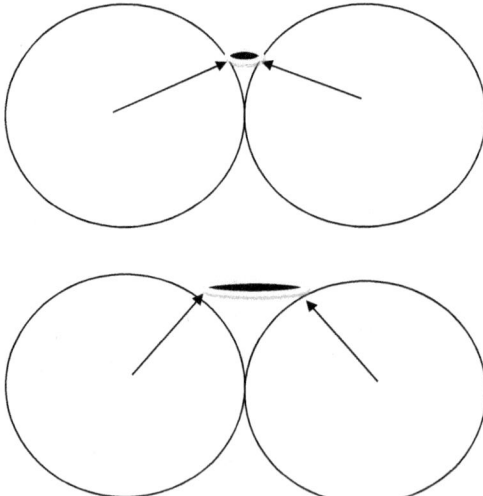

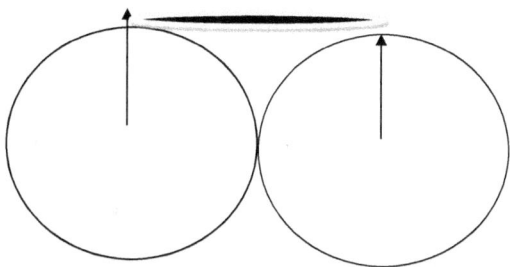

$\bar{h} + R - R\cos(90) = R$

R = distance between the tip of the arrows.

The bubble contains all the matter in the cosmos such as galaxies, and so forth. After stretching along the sides of the doughnut, a hole was formed at the center of the bubble. The hole kept getting bigger until the bubble was transformed into a thin ring.

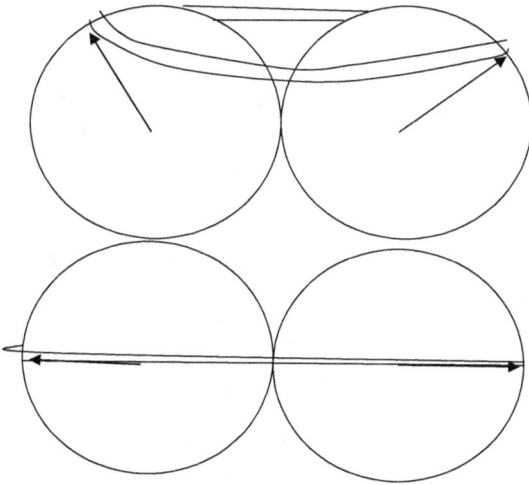

$\bar{h} + R - R\cos(180) = 2R$
2R = distance between the tip of the arrows.

Following the curved path along the sides of the doughnut the cosmic bubble began to contract.

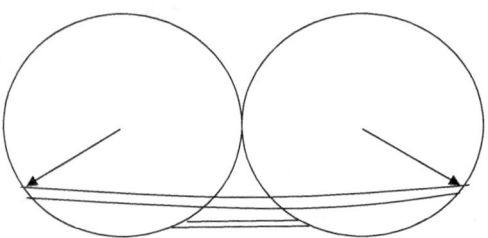

On its way back to the singularity, the cosmic bubble becomes a disk once again, losing its hole in the middle. $\bar{h}+ R - R\cos(270) = R$ is the distance between the tip of the arrows.

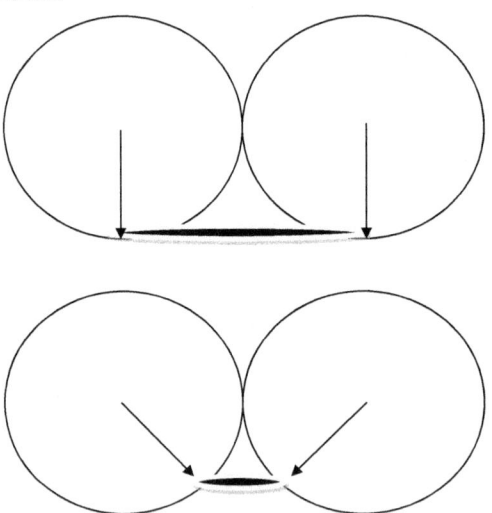

Half of the universal disc is made of matter and the other is anti-matter. When the two separate the universe is created from the singularity, and when the two meet again they are converted to light energy from where a new cycle begins.

The volume that we are able to see in the universe must be doubled if we are to include the second half of the disk that is made of anti-matter.

The Matter and Anti-Matter in the universe meet again at the Big Crunch. After that, a new cycle begins.

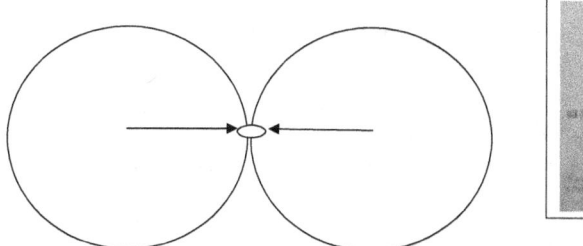
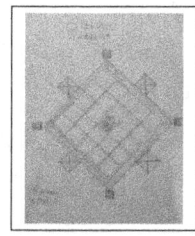

Eventually the cosmic bubble returns to the singularity, exploding in the Big Crunch, becoming a new Big Bang, and doing it all over again and again endlessly.

The equation that describes the distance between the tip of the arrows is = Distance = $\bar{h}$ + R − R cos($\bar{h}(z)$)

The angle is measured from the singularity to the tip of the arrows and R is the diameter of one of the two wheels.

Plotting the graph of that function we get.

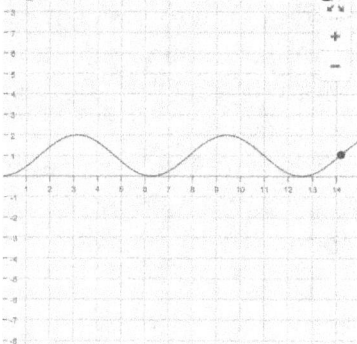

z is angular increments since even angles exist at quantum pieces. z can be 0, 1,2,3,4... and so forth. Only whole numbers which produce the θ needed.

Vertical axis is for distance and the horizontal axis for the angle in radians. When the angle is 180 degrees ($\pi\ radians$), the distance is at its greatest at the peak of the graph. The graph shows about 2.5 cycles of expansion and contraction. The maximum length of the diameter is 2 times the radius of the sphere containing the doughnut.

The total energy of the universe is still zero.
1-1 = 0. This cycle repeats forever since forever.

Particles are vibrating strings shaped like a ring. Particles vibrate Dark Matter ( the Ether ). That is why particles diffract and propagate in space like waves through a medium. The medium in this case is the Ether.

Everything is made of vibrating rings of energy propagating in space, transferring energy to other rings of energy in the cosmos.

The superposition of these waves forms the cosmos. There is an infinite amount of parallel universes. The closer parallel universes are more similar to ours. The farther parallel universes become more and more different the farther you go.

Travelling back in time is moving to a parallel universe. If you go back to 1965, you will not recognize it. It will not be like the 1965 you learned about. It will look like the year 1965 in a parallel universe.

It is this the reality regarding Feynman's Alternate Histories. There is an infinite number of histories and none of them are exactly alike. There is an infinite number of paths in history both past, present, and future.

Everything in the universe is made of Dark Matter. Visible matter is compressed Dark Matter, and empty space is stretched Dark Matter. Dark Matter is made of strings. When the strings are compressed, waves form giving birth to visible matter. When the strings are stretched, no waves form, and Dark Matter becomes invisible.

The doughnut is curved Dark Matter, and the cosmic bubble is the location of all we can see around us. The curved Dark Matter is the result of a gravitational dipole that pushes the bubble in an endless path around and around again.

This theory solves the problem of a universe coming from nothing, and it instead states that there has always been a bubble following the curved path of doughnut without true beginning and with no true end.

Dark Matter is made of Higgs Bosons that is the Ether. Light needs Higgs Bosons to propagate. Although light has no mass it follows the curved path of space time because it propagates through space using the Ether which are Higgs Bosons.

Higgs Bosons also give mass to particles in compressed regions of Dark Matter that becomes the visible and detected matter and energy that we know.

If light were to not require a medium to propagate it would never follow the curved paths of space time but would go straight instead, but that does not happen, so it does need a medium to flow. Space is not empty but is filled with Higgs Bosons that curve space time. Light travels faster in empty space where Dark Matter is stretched instead of compressed.

On the top half of the doughnut, Dark Energy causes Matter and Anti-Matter to expand, and in the bottom half Dark Energy causes Matter and Anti-Matter to attract. This is called a gravitational dipole.

Proving my theory of the gravitational dipole

1. If heavy black holes at the center of galaxies are also surrounded by a doughnut shaped structure where Matter and Anti-Matter are forced to move as described by my theory, then we could say a similar structure could be happening at the center of universe.
2. If the universe is expanding with positive acceleration but **the change in acceleration is negative**. This means that although the universe is expanding ever faster, the negative change in acceleration will cause it to contract in the future along the curved nature of the dipole.
3. If it is proven that heavy black holes are gravitational dipoles. Same could be said for the black hole at the center of the universe.

If a measurement made today of the acceleration of expansion of the universe, and another measurement taken **1000 years from now**, shows that the change in acceleration is negative, then it means that the universe will contract in the future.

Space can't be smaller than the Planck Length which is $1.6 \times 10^{-35} m$. The singularity had a diameter equal to that number. That is why scientists can't unify gravity with the other forces. There simply is not a way for space to be smaller than the Planck Length. The singularity before the Big Bang was at the smallest possible size. The Planck Length.

## Proving my theory of Dark Matter

1 If the path of light curves due to gravity, then it is because it is using the curved medium to propagate. This medium is Higgs Boson, and Higgs Boson is Dark Matter.

2 If light did not use the curved space medium to propagate it would move in a straight line without interacting with the curved structure in space at all.

Let us assume that the arrows of my doughnut shaped universe move around the two wheels ( 2D model ) at a constant angular velocity. These arrows as showed earlier, describe the motion of the edge of the cosmic bubble in its curved path around the doughnut. We get the following:

Distance between the two arrows = $\bar{h}+R - R\cos(\bar{h}(z))$
Where R is the diameter of one of the two spheres. I used R instead of D because R is the radius of the big sphere containing the doughnut. $\theta$ is the angle between the arrows and the singularity.

The derivative of the distance between the two arrows with respect to the angle is:

$\frac{dD}{d\theta} = R\sin(\theta)$, this equation gives the change in the distance between the arrows with respect to the change in the angle $\theta$. It is the velocity of either expansion or contraction.

The derivative of that gives the acceleration of expansion or contraction.

$\frac{\partial^2 D}{\partial \theta^2} = R\cos(\theta)$     Change in acceleration per time = $-R\sin(\theta)$

If under observation it is proven that the change in acceleration of expansion is negative, then the cosmos will likely contract and collapse in the big crunch.

$R^2 = v^2 + a^2$
Knowing the velocity and acceleration of the expansion universe, R can be found.

$(R\sin\theta)^2 + (R\cos\theta)^2 = v^2 + a^2$

That is since $v = \frac{dD}{d\theta}$ and $a = \frac{\partial^2 D}{\partial \theta^2}$

V is the velocity and a the acceleration

Using conservation of energy and assuming that energy is homogeneous throughout the cosmic bubble we can concentrate on Planck's Space.

$$\varrho = \frac{1}{2}m(r\omega)^2 + \frac{GmM}{r+} - \frac{GmM}{r-}$$

Where $\varrho$ is the average energy density of a space the size of Planck's Space. This average density is zero. m is the small mass in that space, G is the gravitational constant, and M is the equivalent mass of the black and white hole in the center of the doughnut. r+ is the curved distance between that mass at the very edge of the cosmic bubble from the white hole in the singularity. r- is the curved distance of the very edge of the cosmic bubble from the black hole in the singularity.

A white hole spits matter pushing matter around the doughnut. On the bottom of the doughnut there is a black hole swallowing matter. Both the black and the white hole are in the singularity contributing to the gravitational dipole.

The r in the $\frac{1}{2}m(r\omega)^2$ is the radius of a wheel in the 2D model of the doughnut.

$$\varrho = \frac{1}{2}mr^2\left(\frac{d\theta}{dt}\right)^2 + \frac{GmM}{r+} - \frac{GmM}{r-}$$

Knowing that the average density is zero:

$$0 = \frac{1}{2}mr^2\left(\frac{d\theta}{dt}\right)^2 + \frac{GmM}{r+} - \frac{GmM}{r-}$$

$$0 - \frac{GmM}{r+} + \frac{GmM}{r-} = \frac{1}{2}mr^2\left(\frac{d\theta}{dt}\right)^2$$

$$\frac{2\left(0 - \frac{GmM}{r+} + \frac{GmM}{r-}\right)}{m} = r^2\left(\frac{d\theta}{dt}\right)^2$$

$$\sqrt{\frac{2\left(0 - \frac{GmM}{r+} + \frac{GmM}{r-}\right)}{m}} = r\frac{d\theta}{dt}$$

$$\sqrt{\frac{2\left(0 - \frac{GmM}{r+} + \frac{GmM}{r-}\right)}{m}} = \frac{dx}{dt}$$

$$\sqrt{\frac{m}{2\left(0 - \frac{GmM}{r+} + \frac{GmM}{r-}\right)}} = \frac{dt}{dx}$$

$$dx\sqrt{\frac{m}{2\left(0 - \frac{GmM}{r+} + \frac{GmM}{r-}\right)}} = dt$$

(r+) + (r- ) = constant. The distance between a point at the edge of the cosmic bubble and the black hole plus the distance between the same point to the white hole is just the circumference of a wheel in the 2D model of the doughnut.

Let's call the circumference C.

(r+) – C = -(r-)   then   C - (r+) = (r-)

So,

*The mass of white hole is negative which leads to the two potentials becoming positive.*

$$dx \sqrt{\frac{m}{2(0-\frac{GmM}{r+}+\frac{GmM}{C-(r+)})}} = dt$$

$$\int_{Planck's\ Length}^{Today's\ length\ for\ r+} dx \sqrt{\frac{m}{2(0+\frac{GmM}{r+}+\frac{GmM}{C-(r+)})}} = T = 13.5 \text{ billion of years}$$

From the above equation we can solve for the (r+) of today's cosmos and verify my gravitational dipole theory. We do know that the current value for (r+) is 13 billion light years away, which is $1.23 \times 10^{26} m$. We also know that 13.5 billion years is $4.09968 \times 10^{17}$ s. Plugging these values in the above equation we get:

$$\int_{1.6 \times 10^{-35} m}^{1.23 \times 10^{26} m} dx \sqrt{\frac{m}{2(0+\frac{GmM}{r+}+\frac{GmM}{C-(r+)})}} = T = 4.09968 \times 10^{17} \text{ s}$$

The variable in the above integral is r+. dx stands for the variable r+. The integral goes from the Planck Length at the beginning of this universal cycle, the Singularity at the Big Bang, to the current value for r+. The integral should equal the current age of the universe in seconds.

Little m is Planck's Mass which is $1.78 \times 10^{-41}$ kg and big M is the mass of the Black-White Hole in the center of the universe which is the total mass of the universe. There are 3 unknowns in the equation which are the value for C and the mass of that Black-White Hole in the center of the universe. If we know C we can then find r-, or either way. $C = 2\pi r$ with r being the radius of one of the circles in the 2D model of the Universe.

### **Volume of the Cosmic Disk**

As the disk shaped cosmic bubble moves along the curved path of the doughnut, its disk becomes thinner and later turning into a thin ring at its maximum diameter expansion. Then it starts getting thicker but the volume decreases until it reaches the singularity with no volume. We can express the thickness of the disk turning into a ring and then into a disk with the equation: $1-1\sin(\theta/2)$

When the arrows representing the path of the cosmic bubble are at 0 degrees, the $1-1\sin(0) = 1$. This means that the disk has no hole in the middle. When the arrows are at 45 degrees, $1-1\sin(45/2) = 0.6173$, which means that 61.73 % of the volume of the disk is made of the cosmic bubble while the remaining 38.27 % is the hole in the middle. The ring keeps getting thinner. When the angle is 180 degrees, 1-

1sin(180/2) = 0, which means that the disk has turned into a thin ring, the thinner it can ever possibly be. This is the only way it can possibly move along the path of the doughnut. The value 0 might seem very radical in saying that the ring is so thin that there is not a ring. In real world that is impossible, but it serves as a mathematical tool to describe a motion of the ring around the doughnut. The Planck Length of space is the minimum it can ever possibly be. We can say that the universe at 180 degrees is as thick as its diameter at the singularity, which is the Planck Length of space (very small). Obviously such an extreme event is really far in the future to happen. We can use the above equation and combine with R-Rcos($\theta$) to find the volume of the cosmic bubble as it moves around the doughnut.

Let's say that the disk has a vertical thickness of a fixed number such as Γ, the symbol representing that constant. By vertical I mean by measuring the thickness using a ruler normal to the plane of the disk. The other thickness stands for a ruler measuring the thickness of the bubble parallel to the plane.

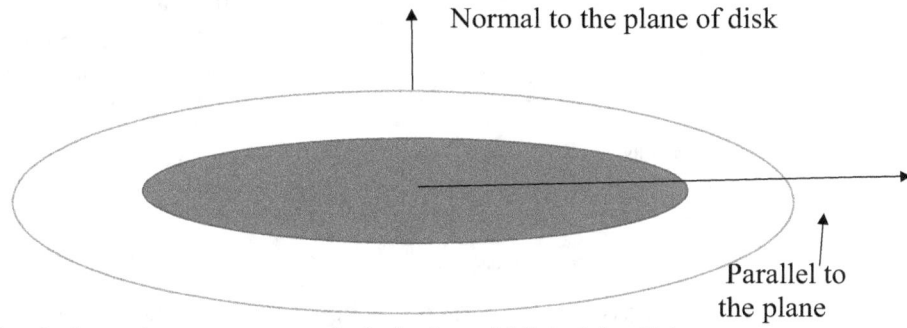

The dark section represents a hole in the middle of the disk.

Let's say that the vertical thickness is constant, but the parallel thickness changes as the disk gets thinner into a ring, or thicker into a fatter ring, or if it has no hole in the middle at all such as in the singularity. The parallel thickness is measured from where the hole begins all the way to the edge of the cosmic bubble ring. A fatter ring is thicker while a thinner ring is thinner.

The area of a circle is $\pi r^2$. We know that the diameter of the disk at any given point is R-Rcos($\theta$) *and that the r*adius of the disc is half the diameter so we get: $\frac{R-R\cos(\theta)}{2}$ = radius.

Given the fact that 1-1sin($\theta$/2) gives the ratio of the region of the disk made up of the bubble compared to the whole disk, and the fact that volume equals areas times thickness we get the following equation for the volume of the cosmic bubble:

$(1-1\sin(\theta/2))\,\Gamma\pi(\frac{R-R\cos(\theta)}{2})\wedge 2 = \text{Volume}.$

2r = R = Radius of the Doughnut Shaped Universe.

When graphed we get the following in radians:

$\Gamma$ can be calculated by using the idea that:

$\frac{d(volume)}{dt}$ = Velocity of expansion

$\frac{d^2(volume)}{(dt)^2}$ = Acceleration of expansion

Using the observation of the current universe, its velocity of expansion can be measured and $\Gamma$ discovered.

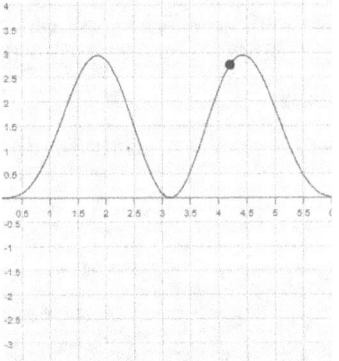

It can be possible to find the $\theta$ representing a given point in the universal cycle from the volume of the universe. If we know the volume of the present universe we can speculate what $\theta$ is by proposing a value for $\Gamma$.

r can be calculated from the equation on the left.

$r = \frac{(r+)}{\theta}$

The vertical axis gives the volume of the cosmic bubble and the horizontal axis the angle in radians. The vertical scale was chosen randomly with the sole purpose to only show changes in volume in the cycle.

The above graph shows 1 cosmic cycle around the doughnut. This shows that although the disk gets thinner, the volume increases until about

1.847radians (105.82 degrees), after that it starts to decrease until the ring disk becomes so thin that it disappears completely in the quantum realm. This happens at around 180 degrees (3.14 radians ). Although the diameter of the cosmic bubble disk gets larger, its parallel thickness gets smaller, which shows that it will eventually cause the volume to decrease to the same value that it has in the singularity. Since in today's universe we see the cosmos expanding, most likely we live between 0 and 105.82 degrees in the cosmic cycle. After 180 degrees, the cosmic bubble begins to contract but the parallel thickness starts to grow, turning the ring into a disk. This thickness causes the volume to increase although the cosmos is contracting. This is shown by the second rise leading to a second peak on the graph. The second peak occurs at around 253.71 degrees ( 4.428 radians). Then the contraction of the cosmic bubble, despite the growing parallel thickness, leads to a decrease in volume until reaching the singularity with 0 volume in the quantum realm at 360 degrees ($2\pi\ radians$). **In the equation 1-1sin($\theta/2$), only degrees between 0 and 360 are allowed or else the graph and data stops making sense as the graph shoots for angles greater than 360 degrees when combined with R-Rcos( $\theta$) with no physical meaning for it.**

The entropy in the universe increases with volume between arrow angles of 0 and 105.82 degrees and decrease with volume between that and 180 degrees. The entropy then increases again with volume between 180 degrees 253.71 degrees to then decrease with volume from that to 360 degrees. One complete revolution of the cosmic bubble around the doughnut gives a change of entropy of exactly zero. The universe is self-contained and fully enclosed.

## **Temperature:**

Treating the cosmic bubble as a gas we can see the relationship:

Temperature = $\frac{\sigma}{Volume}$

Temperature = $(\sigma 2)(Pressure)$

The temperature is related to the volume of the gas. The smaller the volume the higher is the temperature. The $\sigma$ and $\sigma 2$ are constants of proportionality. This relation is the opposite of $\frac{PV}{T}$ from the ideal gas law that shows that the smaller the volume the lower the temperature. In our case here, the smaller the volume the greater is the temperature instead. Also in our case, the higher the pressure the higher is the temperature.

So our general equation should be instead:

$$T = \frac{P}{V}(\sigma)$$ Notice that we are solving for temperature. When the pressure increases, the volume must simultaneous decrease in order for the temperature to get bigger.

The temperature of the universe is the greatest at angles 0 and 180 degrees when its pressure is the highest and its volume the smallest.
The current universe is cooling which shows that the pressure is getting smaller and the volume bigger.

### **Shape of the Cosmic Bubble:**

Bellow is the illustration of the cosmic bubble:

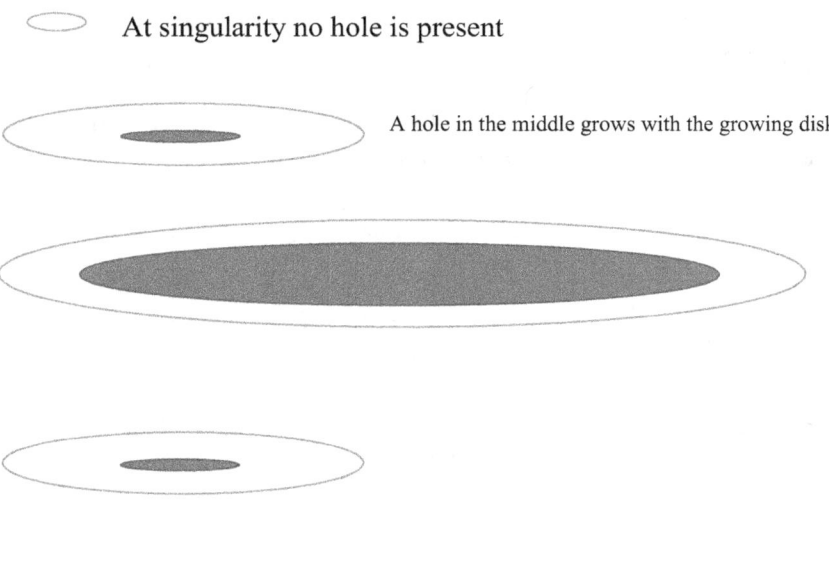

At singularity no hole is present

A hole in the middle grows with the growing disk

Back to Singularity the hole disappears

To prove that the cosmic bubble is a disk we will need to observe a pattern in the cosmic microwave background.

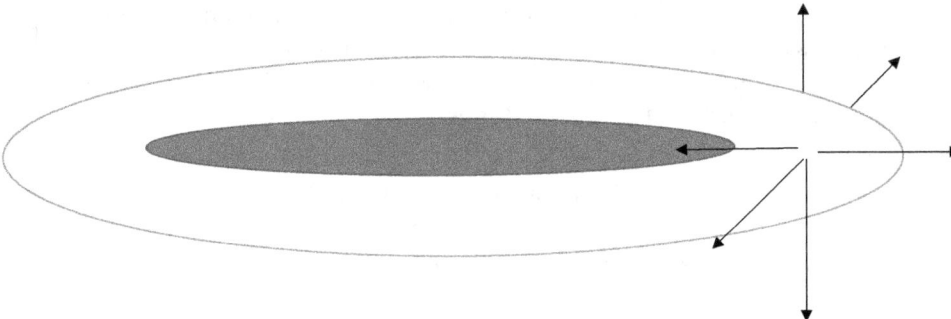

The source of the arrows is a probable location of our galaxy in the cosmic bubble.

From this picture we conclude that because the disk is thinner than it is wide we expect to see more matter in the following directions:

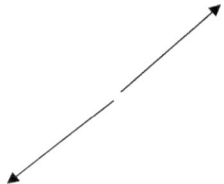

Both locations are 180 degrees from each other. In the cosmic microwave background that would indicate hotter regions.

We would expect the two cooler regions, not as hot as the first case to be in the following directions:

Once again at 180 degrees from each other.

And last but not least, we expect the coolest of all the regions of the cosmos to be the ones looking in the direction of the thickness of the disk:

Also at 180 degrees from each other.

Now we should be able to see a pattern like that in the cosmic microwave background. Since our galaxy and our solar system may be tilted when compared to the plane of the cosmic bubble, what we see will also be tilted. Here is the picture of the cosmic microwave background:

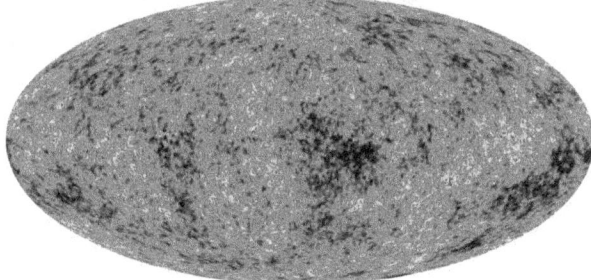

The brighter regions are the hottest.

## Cyclical Universe and the Wheel

$$(1-1\sin(\theta/2)) \; \Gamma\pi(\frac{R-R\cos(\theta)}{2})^{\wedge}2 = \text{Volume}.$$

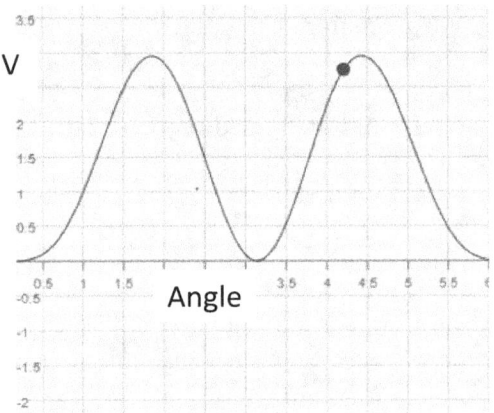

The above equations shows the variation of the Volume of the Universe as the angles in my Dipole Theory changes. The graph represents that function. $\Gamma$ is the parallel thickness which is held constant.

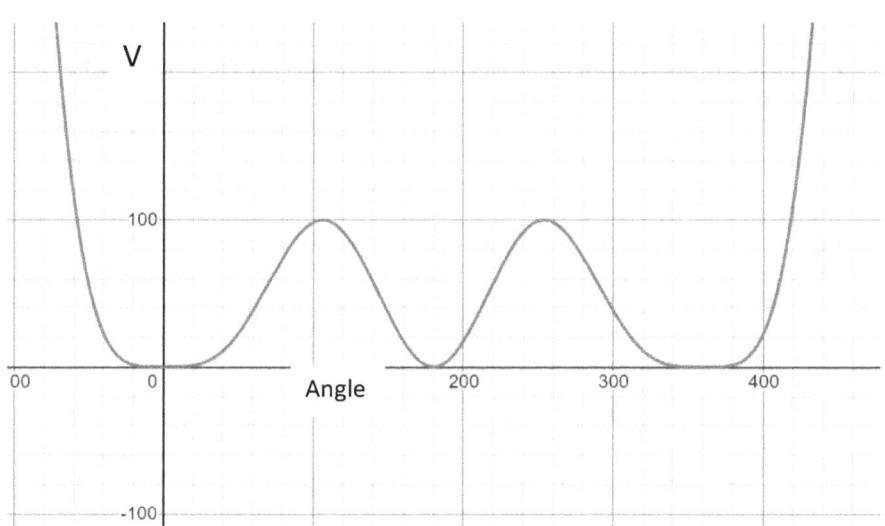

The angles used in the equation on previous page must be between 0 and 360 degrees. Any numbers beyond 360 degrees, or below 0 degrees will cause the graph to shoot up as shown in the graph above. The shooting up has no physical significance.

100 was chosen as the largest volume that the Universe can have as a constant of proportionality with no physics significance.

| Angle | Volume | Angle | Volume |
|---|---|---|---|
| 0 | 0 | | |
| 15 | 0.30730102 | 195 | 9.9568646 |
| 30 | 4.0502649 | 210 | 35.968446 |
| 45 | 16.125776 | 225 | 67.431335 |
| 60 | 38.072427 | 240 | 91.800631 |
| 75 | 65.483323 | 255 | 99.887744 |
| 90 | 89.275964 | 270 | 89.501078 |
| 105 | 99.868705 | 285 | 65.815699 |
| 120 | 92.009366 | 300 | 38.388226 |
| 135 | 67.794665 | 315 | 16.336273 |
| 150 | 36.344381 | 330 | 4.1388495 |
| 165 | 10.193353 | 345 | 0.32186402 |
| 180 | $3.868227 \times 10^{-4}$ | 360 | $7.837042 \times 10^{-9}$ |

Above is the graph for the Volumes with respect to the angles during one complete cycle of oscillation.

## Temperature:

$$\frac{(Z)}{(Volume)} = \text{Temperature}$$

When Volume is equal to Planck's Constant Temperature is equal to: 10^9 Kelvins.

$10^9 Kelvins$ = Highest Temperature the universe can have at the Singularity.

$$\frac{(L)}{Volume} = \text{Pressure}$$

Z and L are constants of proportionality.

Since $Volume = \frac{Z}{Temperature}$

$$\text{Pressure} = \frac{L(Temperature)}{Z}$$

L/Z = Y another constant of proportionality.

## Linear relationship:

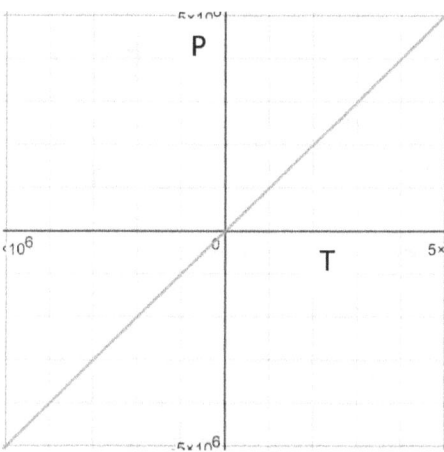

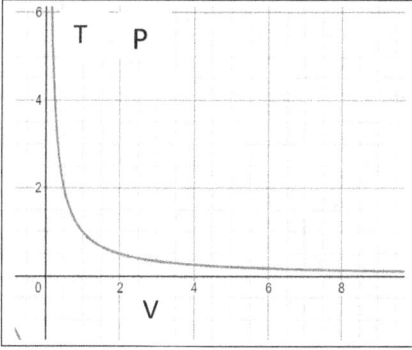

The relationship between Pressure and Temperature is a straight line.

Since $\frac{(Z)}{(Volume)}$ = Temperature

$$\text{Pressure} = \frac{L}{Volume}$$

There is an inverse relationship between (Temperature and Volume), and (Pressure and Volume).

# Distance between the two arrows = $\bar{h}+R - R\cos(\bar{h}(z))$

z is angular increments since even angles exist a quantum pieces. Z can be 0, 1,2,3,4... and so forth. Only whole numbers which produce the $\theta$ needed.

The above equation only gives the variation of the diameter of the ring with respect to the angle.

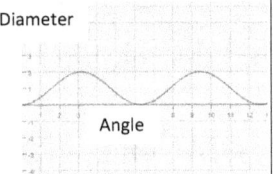

To the left is the function above graphed.

$R^2 = v^2 + a^2$
Knowing the velocity and acceleration of the expansion universe, R can be found.

$(R\sin\theta)^2 + (R\cos\theta)^2 = v^2 + a^2$

That is since $v = \dfrac{dD}{d\theta}$ and $a = \dfrac{\partial^2 D}{\partial \theta^2}$

V is the velocity and a the acceleration

$\dfrac{dD}{d\theta} = R\sin(\theta)$, this equation gives the change in the distance between the arrows with respect to the change in the angle $\theta$. It is the velocity of either expansion or contraction.

The derivative of that gives the acceleration of expansion or contraction.

$\dfrac{\partial^2 D}{\partial \theta^2} = R\cos(\theta)$     Change in acceleration per time = $-R\sin(\theta)$

Step 1

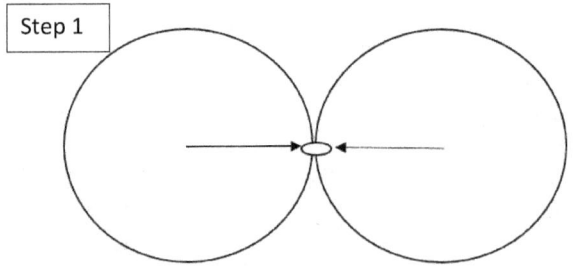

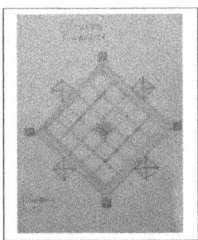

## Doughnut Theory in 2 dimensions:

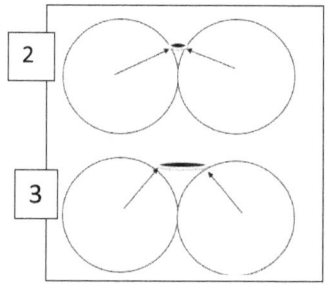

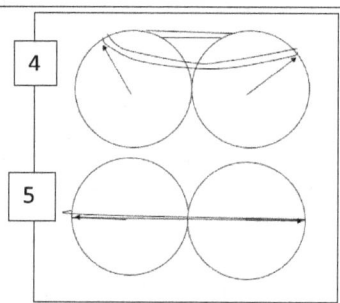

Below is the graph of the change in Volume of the universe from steps 1 – 7.

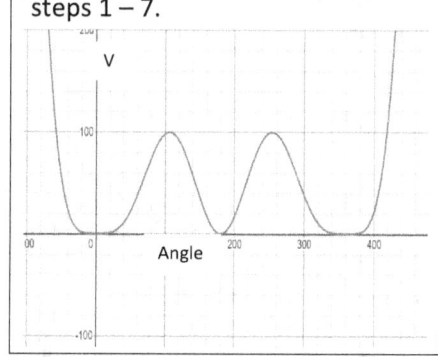

The Equation below can be used to find r+ which is the distance from the Earth to the Big Bang.

$$\int_{1.6\times10^{-35}m}^{1.23\times10^{26}m} dx \sqrt{\frac{m}{2(0-\frac{GmM}{r+}+\frac{GmM}{C-(r+)})}} = T = 4.09968\times10^{17} \text{ s}$$

$$\varrho = \frac{1}{2}m(r\omega)^2 + \frac{GmM}{r+} - \frac{GmM}{r-}$$

The equation of the top was derived from the one the left using the fact that $\varrho$ = zero= Total Energy of the Universe.

The Five Elements that make up the Universe according to Plato:

Land   Water   Air   Fire   Ether

All the universe is made of Energy. All matter is Energy. All Energy are waves oscillating in the strings that make up the Dark Matter.

Gravity is caused by matter pulling the strings in the Dark Matter (Ether) leading to it being attracted to other matter.

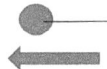 

Charges are caused by the amount of rotation of a particle in Dark Matter. This rotation generates a rotating fluid of light in space leading to + or - charges.

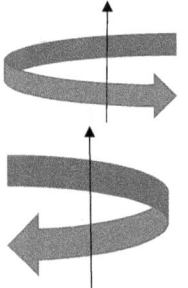

The direction of the spin with respect to a magnetic field generates a charge + or - charge. This spin has nothing to do with the spin in quantum mechanics.

Empty space is made of stretched strings:

Space containing matter is made of string with waves in it.

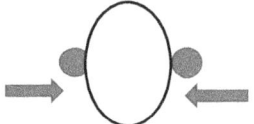

The balls at the end of strings are the Higgs Bosons. When they approach each other gathering, they form a wave in the string between them leading to matter. When the Higgs Bosons move away from each other no waves are formed and that is empty space.

2r = R = Radius of the Doughnut Shaped Universe.

.r = Radius of one wheel in 2D

It can be possible to find the $\theta$ representing a given point in the universal cycle from the volume of the universe. If we know the volume of the present <u>universe</u> we can speculate what $\theta$ is by proposing a value for $\Gamma$.

$$r = \frac{(r+)}{\theta}$$

r can be calculated from the equation on the left.

$(r+) + (r-) =$ constant. The distance between a point at the edge of the cosmic bubble and the black hole plus the distance between the same point to the white hole is just the circumference of a wheel in the 2D model of the doughnut. Let's call the circumference C.

$(r+) - C = -(r-)$ then $C - (r+) = (r-)$

So,

$$dx \sqrt{\frac{m}{2(0 - \frac{GmM}{r+} + \frac{GmM}{C-(r+)})}} = dt$$

$$\int_{1.6 \times 10^{-35} m}^{1.23 \times 10^{26} m} dx \sqrt{\frac{m}{2(0 - \frac{GmM}{r+} + \frac{GmM}{C-(r+)})}} = T = 4.09968 \times 10^{17} \text{ s}$$

From the above equation we can solve for the (r+) of today's cosmos and verify my gravitational dipole theory. We do know that the current value for (r+) is 13 billion light years away, which is $1.23 \times 10^{26}$ m. We also know that 13.5 billion years is $4.09968 \times 10^{17}$ s.

The more compressed a region of space is, with Higgs Bosons clustered in one place, the greater is the Gravitational Pull.

In empty space the Dark Matter (Ether) grid containing the Higgs Boson Spheres has its strings stretched.

In space containing matter, the Higgs Bosons are compressed close together, leading to waves in the Dark Matter String.

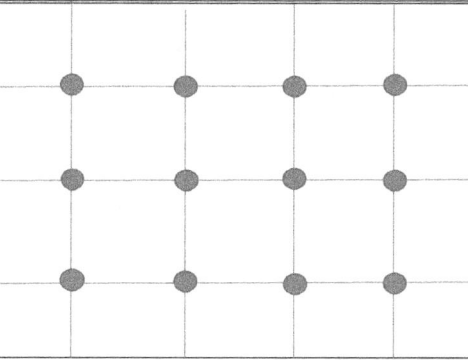
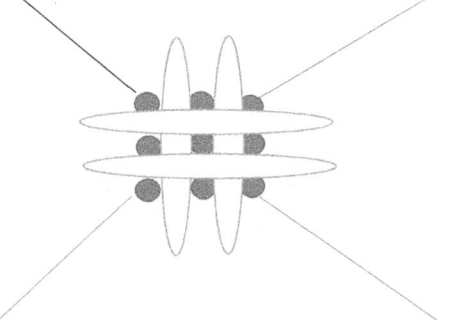

## The Four Forces of Nature:

**Gravity:** Caused by the pull of the strings of Dark Matter.

**Electromagnetic, Weak, and Strong:** Propagation of light waves in the Dark Matter medium.

**Waves:** All matter is energy. All energy is light. All light are waves in the Dark Matter medium.

Everything is made by light since everything is energy.

When a particle or any matter travels through space it uses the Dark Matter of the region it is travelling through as part of their composition. Example is shown below.

Here is empty space with Higgs Boson equally separated from each other.

Here now is a particle travelling through that region of space towards the right:

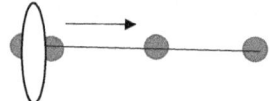

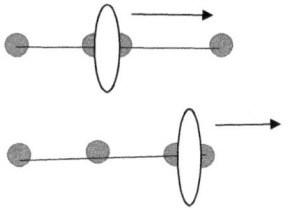

Notice that Dark Matter is used by the particle as its composition. The particle literally goes right through Dark Matter and that is why it is so hard for Dark Matter to be detected.

The strings get compressed in the presence of a particle forming a loop on the particle's location.

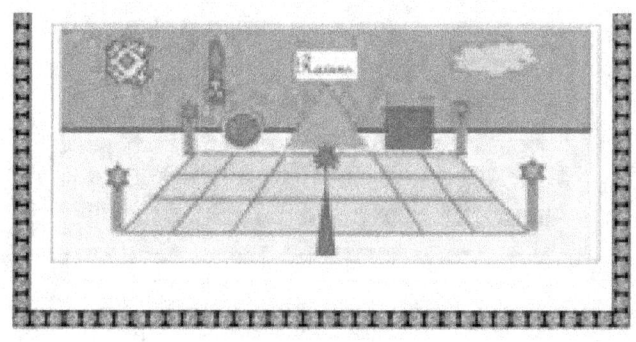

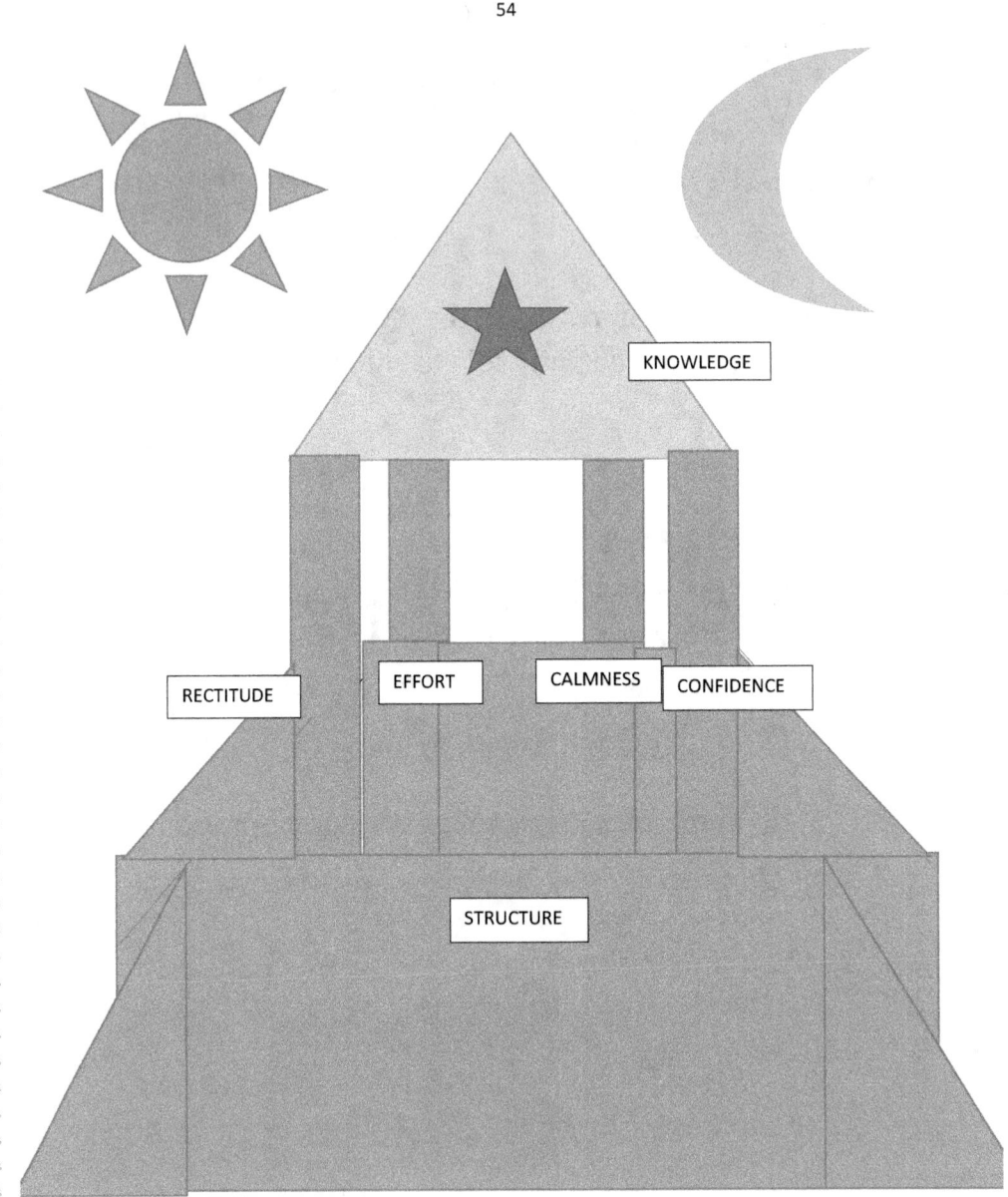

# Rings of Energy

The 3-dimensional axis:

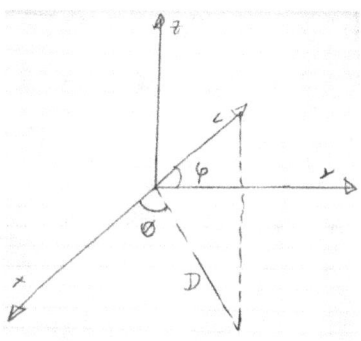

L is the distance between the Particle and the point (0,0,0).

D is the line distance between a point directly below or above the Particle on the x,y plane to the center (0,0,0).

$\theta$ is the angle between the x axis and the line D.

$\varphi$ is the angle between the x,y plane and th Particle.

---

Using these 3 coordinates, we get the following for each coordinate:

| X | Y | Z |
|---|---|---|
| $L(\cos(\varphi))(\cos(\theta))$ | $L(\cos(\varphi))(\sin(\theta))$ | $L(\sin(\varphi))$ |

These come from the following:

$X^2 + Y^2 = D^2$

$D(\cos(\theta)) = X$

$D(\sin(\theta)) = Y$

$D^2 + Z^2 = L^2$

$L\sin(\varphi) = Z$

$L\cos(\varphi) = D$

$X^2 + Y^2 = L^2 - Z^2$

$X^2 + Y^2 + Z^2 = L^2$

Let us say that the Zyn Particle is moving in a helix pattern in the x direction. Looking at it from above we see the helix and the Particle going in the x direction. Looking at the motion by being in front of it, we only see a circle in the y,z plane:

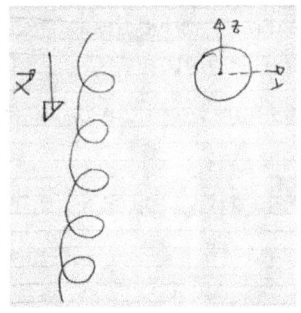

We get the following equation from the coordinates from this motion:

X = (speed)(time)

Y = $L\cos(\varphi)$

Z = $L\sin(\varphi)$

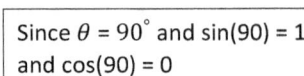

Since $\theta = 90°$ and $\sin(90) = 1$ and $\cos(90) = 0$

Let us say that the particle rotates in the y,z plane while moving in the x direction at angular Frequency $\omega_\varphi$. We then get the following:

X = $v_x t$

Y = $L\cos(\omega_\varphi t)$

Z = $L\sin(\omega_\varphi t)$

Let us say that while the Particle moves in a helix pattern in the x direction, in itself oscillates sinusoidally with Frequency $\omega_L$ which means that the value of L changes according to:

L = $L_o \sin(\omega_L t)$

The new equation for Y and Z then becomes:

Y = $L_o \sin(\omega_L t)(\cos(\omega_\varphi t))$

Z = $L_o (\sin(\omega_L t))(\sin(\omega_\varphi t))$

Let us say that this Particle besides moving in a helix in a sinusoidal pattern it is also moving through a tube that forms a perfect circle around a center point. The distance between this center point and the center of the tube is R.

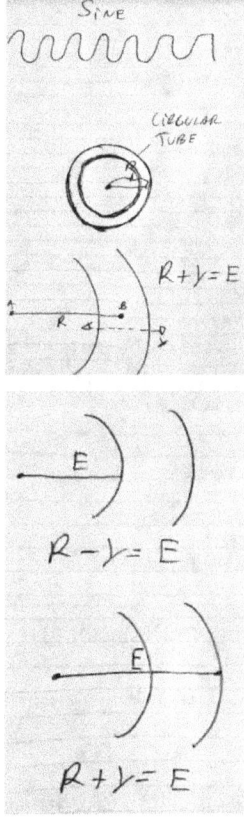

The distance between the center of the circle and the inner edge of the tube is R plus the maximum negative value of y. The distance between the center of the circle and the outer edge of the tube is R plus the maximum positive value of y. The equation for the distance between the center of the circle and the horizontal component of the location of the Particle inside the tube is E = R + Y

E is the horizontal component of the distance between the Particle and the center of the circle along the x,y plane where E is located.

E = R + Y

The x and y in the x,y plane of E is not the same X and Y of the Particle inside the tube.

Let us say that the Particle moves around the tube with Frequency $\omega_E$

We get the following:

$E_x = E\cos(\xi)$   With $\xi$ the angle between the line E and the positive x axis of the x,y plane in E.
$E_y = E\sin(\xi)$

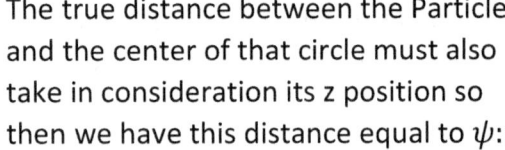

$E_x = [R + L_0\sin(\omega_L t)(\cos(\omega_\varphi t))]\cos(\omega_\xi t)$

$E_y = [R + L_0\sin(\omega_L t)(\cos(\omega_\varphi t))]\sin(\omega_\xi t)$

The true distance between the Particle and the center of that circle must also take in consideration its z position so then we have this distance equal to $\psi$:

$E^2 + Z^2 = \psi^2$   Unlike x and y, the z coordinate of the particle inside the tube is the same z of the particle above and below the plane where E is located.

With

$Z = L_0(\sin(\omega_L t))(\sin(\omega_\varphi t))$

The equation for $\psi$ is then using Pythagorean Theorem:

$$\psi = \sqrt{[R + (L_0\sin(\omega_L t))(\cos(\omega_\varphi t))]^2 + [L_0(\sin(\omega_L t))(\sin(\omega_\varphi t))]^2}$$

Since sine square plus cosine square equals 1 the $\cos(\omega_\xi t)$ and $\sin(\omega_\xi t)$ are gone is the equation above.

If v is equal to the speed of an electron = $2.16 \times 10^6 m/s$ and R the radius of a Hydrogen Atom = $5.29 \times 10^{-11} m$.

It takes the Particle $1.54 \times 10^{-16} s$ to make one revolution.

Which makes its $\omega_\xi$ = 4.08x10$^{16}$ radians/second

$\omega_L$ = Is the Frequency of the Particle oscillating sinusoidally which is equivalent to its Energy.

$\omega_\varphi$ = is the Frequency of the Particle rotating in a helix. The faster the rotation the more charge the Particle has. Clockwise and Counterclockwise motion leads to + or – charges. No rotation in helix is equivalent to a neutral Particle.

Particles can move at any pathways besides a circular tube:

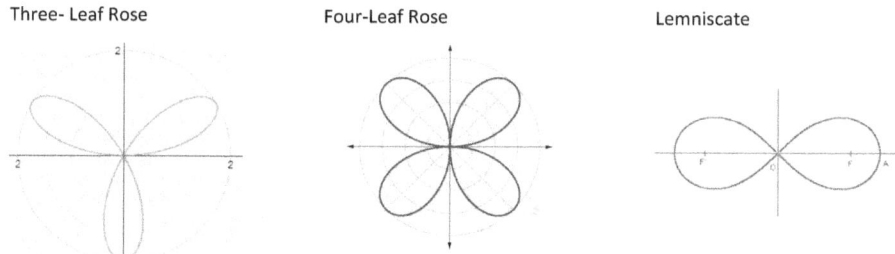

Three- Leaf Rose    Four-Leaf Rose    Lemniscate

These equations can be very complex and be able to give an in depth investigation of the probabilistic wave cloud of Electrons in orbit of Atoms. **Maybe the motion of Electrons and Particles in general are not at random inside probability clouds, but do follow a repetitive pattern in which it is so complex that we can only grasp it by using probabilities.**

**Investigation of the Equation:**

The two potentials are shown positive at the bottom of page since the white hole that spits out matter in the Big Bang has negative mass which combined to the negative in front becomes positive.

$$\sqrt{\frac{2(0 - \frac{GmM}{r+} + \frac{GmM}{r-})}{m}} = \frac{dx}{dt}$$

In performing an investigation of the equation above, the term $\frac{dx}{dt}$ is the speed at which the arrows rotate in the doughnut.

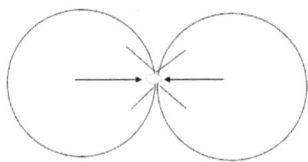

It can be proven that the arrows move in roughly the same speed except at the very beginning and at the very end of each cycle. The limit at the very beginning is that the universe can not be smaller than the Planck's Length which means that the Planck's Length is the size of the universe at time zero.

Ignoring the constants in the following equation and using unit 10 for the circumference of the circles of the 2D model of the doughnut, we get the following speeds as the arrows move in a complete circle:

$$\sqrt{\left(\frac{1}{x}\right) + \left(\frac{1}{10-x}\right)} = \frac{dx}{dt}$$

With x = 1

$$\sqrt{\left(\frac{1}{1}\right) + \left(\frac{1}{10-1}\right)} = 1.054$$

$$\sqrt{\left(\frac{1}{2}\right)+\left(\frac{1}{10-2}\right)}=0.791$$

$$\sqrt{\left(\frac{1}{3}\right)+\left(\frac{1}{10-3}\right)}=0.690$$

$$\sqrt{\left(\frac{1}{4}\right)+\left(\frac{1}{10-4}\right)}=0.6455$$

$$\sqrt{\left(\frac{1}{5}\right)+\left(\frac{1}{10-5}\right)}=0.6325$$

$$\sqrt{\left(\frac{1}{6}\right)+\left(\frac{1}{10-6}\right)}=0.645$$

$$\sqrt{\left(\frac{1}{7}\right)+\left(\frac{1}{10-7}\right)}=0.690$$

$$\sqrt{\left(\frac{1}{8}\right)+\left(\frac{1}{10-8}\right)}=0.791$$

$$\sqrt{\left(\frac{1}{9}\right)+\left(\frac{1}{10-9}\right)}=1.054$$

Notice that between 1/10 and 9/10 of one revolution the speed of expansion is roughly somewhere between 0.7 and 0.6 which means that the arrows can be assumed to move at somewhat constant speed.

The speed that the arrows move, however, change dramatically as the bubble universe is near the beginning and at the end of a cycle. For example:

$$\sqrt{\left(\frac{1}{0.001}\right) + \left(\frac{1}{10 - 0.001}\right)} = 31.62$$

$$\sqrt{\left(\frac{1}{9.999}\right) + \left(\frac{1}{10 - 9.999}\right)} = 31.62$$

Notice that the number begins to increase dramatically as we approach zero or 10. As we approach smaller ratios than 1/10 of a cycle or bigger ratios than 9/10 but smaller than 10/10.

This is the proof for inflation theory that states that at the very beginning the universe had a rapid expansion beyond the speed of light and then reached a somewhat constant speed of expansion. Between 1/10 and 9/10 of a cycle the speed of expansion and contraction is roughly the same but small changes lead to acceleration. Knowing the current speed of expansion and the acceleration we can find R the diameter of each of the two spheres of the 2D model of the doughnut using the following equation:

| $R^2 = v^2 + a^2$ Knowing the velocity and acceleration of the expansion universe, R can be found. | $(R\sin\theta)^2 + (R\cos\theta)^2 = v^2 + a^2$ <br> That is since $v = \frac{dD}{d\theta}$ and $a = \frac{\partial^2 D}{\partial \theta^2}$ <br> V is the velocity and a the acceleration |
|---|---|

In knowing R and r+ which is the distance between us and the point of the Big Bang we can find the angle at which the arrows are making with the singularity at the current moment using the following equation:

$$r = \frac{(r+)}{\theta}$$

Where r is half the length of R.

By knowing the angle $\theta$ that the arrows are making with the singularity right now and by also knowing the volume of the current universe we can find the thickness of the flat universe using the following equation:

$$(1-1\sin(\theta/2))\, \Gamma\pi\left(\frac{R-R\cos(\theta)}{2}\right)^{\wedge}2 = \text{Volume}.$$

Where the thickness is represented by the letter $\Gamma$

Even though the arrows move in roughly the same speed between 1/10 and 9/10 of a cycle, they can generate a pattern in the expansion of the universe that leads to an apparent positive or negative acceleration of the rate of change of the distance between the tip of the two arrows which leads to the following graph from page 49 for the change in volume for one complete cycle:

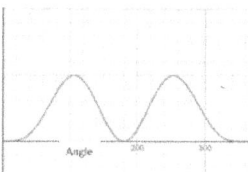

This implies that observations of the current universe must be performed to match the numbers correctly to be able to prove all of this.

Space is made of the Ether which is composed of Higgs Boson Spheres connected to each other by strings of Electric and Magnetic Field:

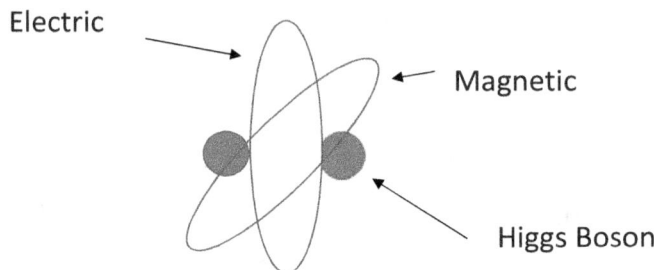

Between each of the two Higgs Bosons there are two loops of strings. One is Electric and the other Magnetic.

The pulling of strings generates Gravity. In regions of space where there are a lot of mass the Higgs Bosons are closer together almost touching. This causes a more pulling of strings and thus more Gravity.

When the Higgs Bosons do touch each other, the distance between their centers is the Planck's Length and space is the densest.

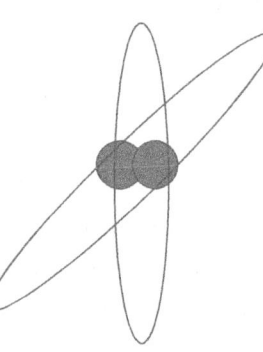

When the Higgs Bosons are closer together the strings have a higher Amplitude.

A single Photon is made of two Higgs Bosons connected to each other by an Electric and Magnetic String at the Fundamental Level:

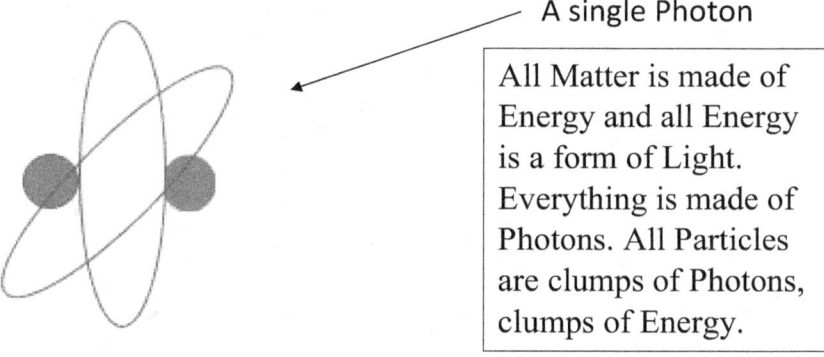

A single Photon

All Matter is made of Energy and all Energy is a form of Light. Everything is made of Photons. All Particles are clumps of Photons, clumps of Energy.

When Photons propagate in space they lead to the Electromagnetic, Weak, and Strong Forces of Nature.

Throughout space the Electric and Magnetic Fields Alternate in the Oscillation:

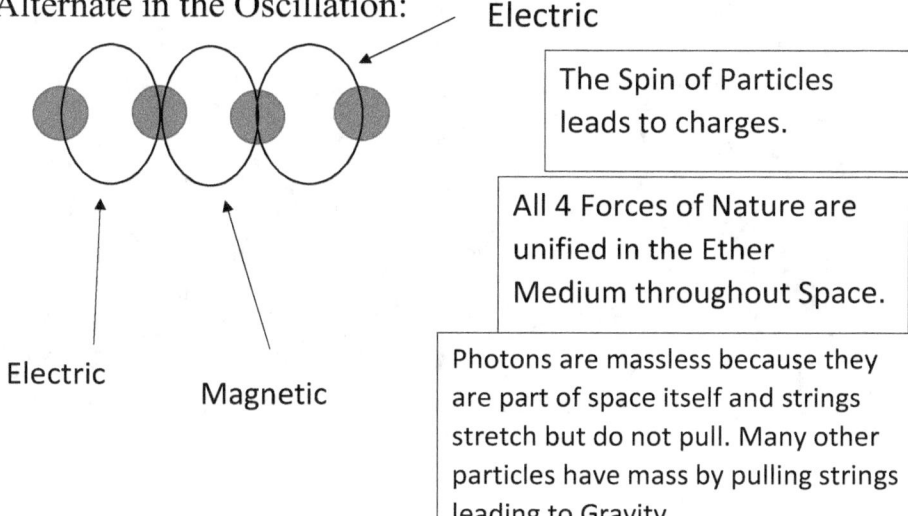

Electric

Magnetic

The Spin of Particles leads to charges.

All 4 Forces of Nature are unified in the Ether Medium throughout Space.

Photons are massless because they are part of space itself and strings stretch but do not pull. Many other particles have mass by pulling strings leading to Gravity.

## Pictures that Inspire

### Songycraype

### Zeno the dreamer

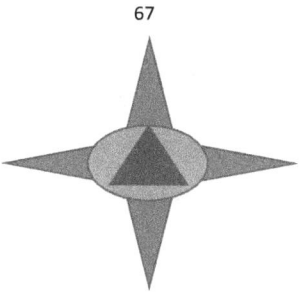

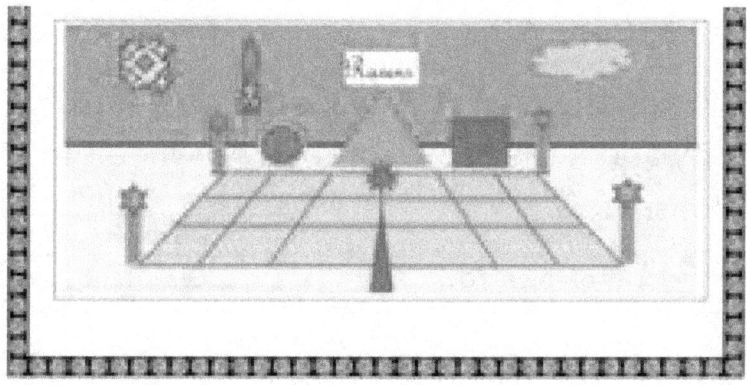

# Discover and Learn

## Steps into proving this Theory:

Find the distance with a telescope from Earth to the very edge of the universe which is the farthest back in time that humans can see. This distance in meters corresponds to r+.

Knowing that r+ = $\theta R$

With $\theta$ being the angle between the tip of arrows and the singularity and R the radius of the two wheels in two dimensions.

The current velocity of the arrows = $\sqrt{\dfrac{2(0 - \dfrac{GmM}{r+} + \dfrac{GmM}{r-})}{m}}$

With m being Planck's Mass and M the Total Mass of the Universe and:

r- = C − r+

The value of C that makes the integral below work is used to find R.

$\int_{1.6 \times 10^{-35} m}^{1.23 \times 10^{26} m} dx \sqrt{\dfrac{m}{2(0 - \dfrac{GmM}{r+} + \dfrac{GmM}{C-(r+)})}} = T = 4.09968 \times 10^{17}$ s

Using the Integral above we find C with the variable dx being r+

Since C is the Circumference of the circles in two dimensions

$C = 2\pi R$

R then is $R = \dfrac{C}{2\pi}$

Knowing R we can find the current angle $\theta$:

r+ = $\theta R$

$\theta = \dfrac{r+}{R}$

Knowing $\theta$, R, and the Volume of the Universe we can find the Thickness $\Gamma$ of this Flat Universe using:

$(1-1\sin(\theta/2))\ \Gamma\pi(\dfrac{R-R\cos(\theta)}{2})\wedge 2 =$ Volume.

The Volume of the Visible Matter in the universe must be multiplied by two in order to also include the Anti-Matter side of the Cosmic Disc Bubble.

The Volume of the Visible Universe times 2 equals the Volume of the Entire Universe.

If all these numbers match and fit exactly into describing the current cosmos both past and future, then the Theory is proved correct.

The Universe is Symmetrical, and it has beginning at the Big Bang, middle, and an end at the Big Crunch.

If we know how long it takes for half of the Universe's Final Age we can multiply that by two to get the Life Expectancy of the Universe.

Since $C = 2\pi R$

Half of the Universe's Cycle will be $C/2 = \pi R$ = value of variable r+

Integrating within these bounds:

$$(2)\left(\int_{1.6 \times 10^{-35} m}^{\pi R} \frac{m}{m} \, dx \sqrt{\frac{m}{2\left(0 + \frac{GmM}{r+} + \frac{GmM}{C-(r+)}\right)}}\right) = \text{Life Expectancy in s}$$

The Life Expectancy of the Universe is how long it takes for One Complete Cycle which is twice the time that takes for half a cycle.

**The UNIVERSE IS MECHANICAL WITHIN HUMAN REACH OF UNDERSTANDING IT**

## About the Author

    I, Diogo Franklin de Souza, was born in the city of Rio de Janeiro, Brazil in August 20, 1986. I moved to Dallas, Texas when I was 11 years old. I write stories since I was 9 years old. My books tend to contain short summaries of the most important things I find about life, morality, religion, philosophy, and science. Like I say, everything is part of a whole system, and this is also for everything I do and write. I always wanted to have all the most important knowledge in only a few short books. That is why I write, and that is my inspiration for short summaries. I hope this book brings some inspiration also for the readers, because that really is the purpose of my work. Read it and take from it, pieces of gold for you that can be useful in your life. Enjoy….

www.ingramcontent.com/pod-product-compliance
Lightning Source LLC
Chambersburg PA
CBHW050015230526
45470CB00003B/978